Personal Wat[ercraft]

Text and illustrations by Jak[e]
Additional material by courtesy o[f]
Thanks to Mark Rowley of Sowester L[td]

Buoyage	*opposite*
Welcome to the water	2
Personal Watercraft Course Syllabus	3
Types of PW	4
Personal gear	5
Layout and controls	6
Nautical Terms	7
Essential Gear	8
Pre-departure checks	9
Pre-launch checks	10
Launching	11
Getting underway	12
Basic steering and trim	13
Capsize and re-boarding	14
Rules of the road	15
Courtesy and common sense	16
Collision avoidance	17
Crossing	18
Overtaking	19
Speed	19
Waterskiing	20
Inshore waters	21
Further offshore	22
Tides and tidal streams	24
Local regulations	26
Courtesy to other water users	27
Sources of weather information	28
Knots, anchoring and towing	29
Recovery from water	30
Safety and emergencies	31
Unacceptable risks	32
Aftercare and maintenance	33
Check list	*rear cover*

© 1998 Royal Yachting Association.

All rights reserved. No part of this publication may be reproduced, stored in a retrieval system, or transmitted, in any form or by any means, electronic, mechanical, photocopying, recording or otherwise, without the prior permission of the publisher.

WELCOME TO THE WATER

Whether you are a first time personal watercraft - PW- owner or an experienced rider, this RYA guide will help you get more enjoyment from your craft. Learn to use your PW responsibly so that you, your passengers and other water users stay safe. Read it and find how to protect your craft - it's a major investment, don't trash it.
Safe, courteous riding is the way to go.
You are a water user, learn the rules and follow them.

PERSONAL WATERCRAFT COURSE SYLLABUS

The one day personal watercraft course covers the topics set out below and is usually taught in the order shown, however, the order may be changed due to local conditions or the weather on the day.

INSTRUCTION ASHORE
INTRODUCTION
Layout of a PW; controls; propulsion and steering system; fuel and oil; stowage compartment
Personal equipment, wet suit/dry suit, personal buoyancy, head and eye protection
Pre launch checks
Essential safety information. Kill-cord, safe speed, local hazards.
COLLISION AVOIDANCE
Rules of the road applied to PW, including:
Lookout
Safe speed
Priorities between different classes of vessel.
Overtaking, crossing and end-on approach rules
Local rules, speed limits, prohibited areas.
ORIENTATION AT SEA
Charts, scales, direction and distance
Representation of land, shallows and deep water. Buoys, lateral and cardinal.
avoiding shipping channels, special buoyed areas for water skiing etc.
tides, high and low water and tidal streams.
WEATHER, SAFETY, COURTESY TO OTHER WATER USERS
Sources and significance of weather forecasts.
Lee and weather shores
Safety and emergency equipment
Courtesy to other water users
Avoiding pollution and disturbance and damage to wildlife habitats.

INSTRUCTION AFLOAT
LAUNCHING AND FAMILIARISATION
Launching from a trailer; board in shallow water and start engine.
Control at slow speed. Balance and trim.
Falling off and reboarding. Capsize and righting.
Control at speed. Stopping distances.
ORIENTATION
Following a planned route, identifying buoys and marks.
COLLISION AVOIDANCE
Recognising potential collision situations and taking correct avoiding action.
PW CONTROL AT SPEED
Slalom exercise
EMERGENCIES
Towing a PW - knots, bowline, round turn and two half hitches.
On completion of this practical exercise, recover PW from water and prepare for trailing and storage.

TYPES OF PW

Designed by a motorcycle-mad Texan, a Personal Watercraft is best described as a boat powered by a waterjet but with an enclosed hull; sometimes called wetbikes, or Jetskis although Kawasaki copyrighted the latter as a trade name.

Stand on or kneel on

Multiple seater

How it works

Water is drawn into the chamber and squirted out again under pressure, rather like placing a thumb over a garden hose. The direction of the jet is changed by the handlebars, so the craft can be steered.

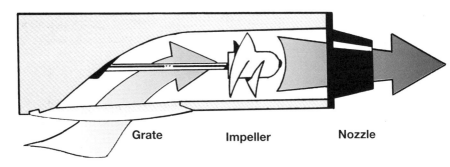

Grate   Impeller   Nozzle

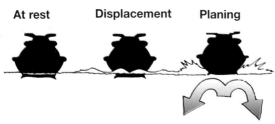

At rest   Displacement   Planing

A PW has a planing hull. The flat bottom and spray rails provide lift and the PW skims the surface. Below planing speed, they are displacement boats, and create a sizeable wash.

PERSONAL GEAR

Powering across the water at speeds of up to 60 mph requires personal protection from sun, spray, or cold. An essential is a personal buoyancy aid which should be a minimum of 50 Newtons and, ideally, a specialist PW buoyancy aid designed to be impact resistant. Many will have a strong point for the attachment of the kill-cord.

Wet suit - many types and styles are available and recommended. Look for a snug fit preferably with an alloy zip. The short sleeved types are only for high summer.

Foot protection - needed for a good grip on the footrests. Useful during launch and recovery.

Whistle - to attract attention.

Kill cord - to stop the engine if you fall off. These are often coded, so act as a security device as well. Make sure it has a strong connection to your clothing or wrist.

Gloves - to help you grip wet controls.

Goggles - spray stings and the water reflects a lot of glare. Wear special PW goggles, or place goggles over your shades or prescription glasses.

A **buoyancy aid** will help keep you afloat.

Dry suits are ideal off season or in cooler climates, but need an under layer. Don't rely on the suit's buoyancy - wear a buoyancy aid or life jacket as well. Squeeze the excess air out before you go afloat or the trapped air may cause you to float ankles up.

Sun Block. A few minutes on a PW with reflected sun and windburn are equivalent to several hours on the beach.

LAYOUT AND CONTROLS

Before using your PW for the first time, read the manual and familiarise yourself with the controls and routine maintenance schedule. Some manufacturers supply a video. Remember, Americans do things slightly differently, especially when it comes to navigation marks.

PW's have a watertight compartment for stowing gear.

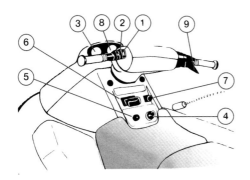
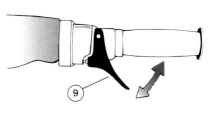
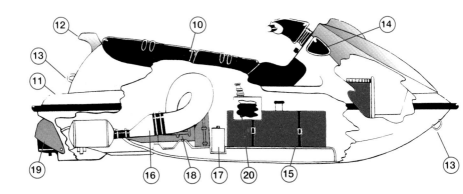

Engine - usually a high revving two stroke petrol unit with two or three cylinders which can develop over 100hp and give speeds of up to 70mph. There is no gear box, clutch (only a few models have neutral or reverse) or brakes. On most units the perol and oil are in separate tanks and are mixed automatically.

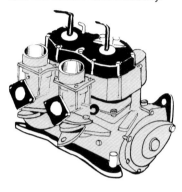

NAUTICAL TERMS

The PW is a boat so there are some simple terms to describe parts of the hull, useful for understanding the collision regulations.

Port and Starboard refer to the boat as seen from the helm looking forwards. Starboard is always the right side of the craft, port is always the left.

Anything behind the helm is usually described as aft.

Typical controls - may vary on different makes and models
1 Stop/start button
2 Variable Trim System switch (optional)
3 Speedometer - a guide only
4 Kill cord attachment - stops the engine if you fall off
5 Choke
6 Reverse gear (optional)
7 Fuel switch
8 Fuel gauge
9 Throttle
10 Seat (removable to get to engine underneath)
11 Tell-tale outlet for cooling system
12 Grab handle
13 Rope securing points
14 Mirrors
15 Fuel tank
16 Exhaust
17 Battery
18 Engine
19 Bucket reverser
20 Oil tank

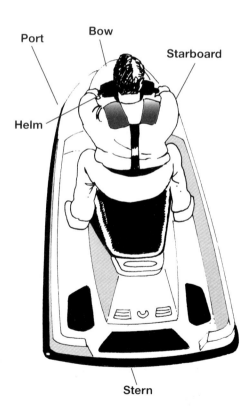

ESSENTIAL GEAR

Fire extinguisher (2kg)

Flare pack *recommended minimum is one pinpoint red and two orange smoke*

Mini-flare packs *fit onto the buoyancy aid and are useful for attracting attention if you are separated from the craft.*

Knife

Torch *for signalling*

5 metres of 8mm nylon rope *for towing, securing to a mooring, or anchoring.*

Small first aid kit

Tool kit and spare spark plugs

Small grapnel anchor folds flat

Documentation *some local authorities insist on a permit and proof of insurance, and have patrol boats that will check.*

Cash *you may end up some way from your car*

Food and drink

PRE-DEPARTURE CHECKS

Is your trailer legal? Remember the speed limit; it's 50mph (60 on motorways)

Check trailer tyres, they work at different pressures to a car.

Is PW use permitted?

Is there a slipway?

Do you need a license?

Are there any restrictions?

Is the PW fully fuelled?

Do you have enough fuel for a full day out?

Is the forecast OK? When is high and low water? It'll help with launching to know.

Check controls are free, and throttle is smooth. Salt water is very corrosive.

Check hull for damage.

Is all the gear aboard?

Is the battery OK? PW left standing may need a battery boost. Take the battery off the machine when charging, as it might damage the systems.

TRAILER TIPS

- Keep the nose down. Bring the strap downwards before it goes onto the winch. This will give added security to the load by holding it more firmly onto the trailer.

- Before attaching the light-board to the trailer, place it alongside the driver's door and work the foot-pedals and indicators. You can see at once if you have a problem with any of the lights. Wrap spare cable around the trailer to stop it dragging on the ground and shorting out.

- Don't put straps across the machine. This can distort the hull. Use the ring provided at the back.

In your tow vehicle carry:
Spare fuel and oil
Spare kill cord
Jump leads

PRE-LAUNCH CHECKS

Avoid hassle by checking local conditions before taking to the water. Visit the local harbour office, ask the beach warden or the local boat or PW clubs.

Check local bye-laws. Is a permit required? Is there a charge for the slipway?
Some authorities require a number to be displayed on a PW.
Park the car well above the high water mark.

Check the controls and hull again in case of damage in transit.

Remove straps and number board (People do forget!)
Make sure the bung is in, or your trip will be a short one
Check engine compartment for any oil or fuel leaks. Mop up any spillage and dispose of it properly.

Use a torch to check the impeller. Make sure no road debris has found its way into the chamber.
Test start and stop the engine once on the main switch and again using the

kill-cord. (Note - in built up areas several PW's test running out of the water can cause a noise nuisance. Start the PW before setting out, and start when immersed, but before leaving the trailer.)

Study the launch site. Watch other users to see which way the tide is flowing, how steep is the slipway? Where can the craft be secured after launch? Other water users or harbour staff will be able to offer advice - sometimes gained the hard way.

LAUNCHING

If you're new to using a trailer, have a practice first. Reversing is the area where most people have trouble as the trailer steers the opposite way to the wheel. Practising will save time and frustration on the slipway which may get busy at weekends.

Slipways or hards vary

Some are excellent at all states of the tide, some stop abruptly below the water, others are cratered, some end in thick mud. Try and get some local knowledge if you are unsure.

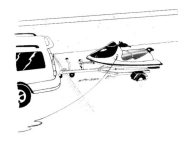

Single handed launch

Long line attached to craft. Security kill-cord with owner, so no-one can steal craft whilst he's parking the car. PW secured alongside jetty or wall, fenders can be used to protect the hull. The underwater hull may get damaged by waves from passing boats if left on the concrete slip. Always secure craft, rising tide may lift a beached one so it drifts away.

Two handed launch

Rider mounted, buoyancy aid done up, lanyard connected, pre-launch checks completed. PW can be pushed off into deeper water. Try and get the machine pointing in roughly the right direction. Use reverse, if you have it, to get off the trailer.

Shallow slipway

On shallow slips car exhaust may be immersed before PW is properly afloat. To avoid this, attach a long line to the trailer and push it out into deeper water with rider mounted.

Trailer is then recovered by securing the rope to the hitch.

And off you go

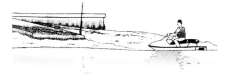

Rider starts engine and motors away slowly.

TRAILER TIP
- When you arrive have a ten minute break before launching to allow the bearings in the trailer to cool down. Hot bearings suck the cold water past the rubber seals and into the bearings, shortening their lives.

GETTING UNDERWAY

Climbing aboard
PWs are quite tippy, so when boarding from a jetty, try and place your weight across the machine.

In shallow water, board from the rear, keeping your centre of gravity low.

From a beach try not to operate engine in less than 3ft of water then motor into deeper water slowly so pump does not pick up sand or gravel and sling it at other beach-users. Stones will make short work of your impeller and require expensive repairs.

In shallow water, watch out for weed, proceed slowly, varying the throttle to avoid entanglement.

Make sure you haven't left a mooring line trailing in the water.

Watch out for the swimming area, often marked with buoys or flags on the beach. Get well away before opening up. Waves will hide bobbing heads from sight. Beaches with designated swimming areas will have a buoyed channel to take you safely offshore.

Keep an eye on your fuel - use 1/3rd going out, 1/3rd coming back, keep 1/3rd in reserve.

Harbours have speed limits. Stick to them. 1 knot is slightly more than 1mph - 6 knots = 7mph. Don't trust your speedometer (or log), which is unreliable at low speeds. Go no faster than a fast walking speed in a harbour area.

BASIC STEERING AND TRIM

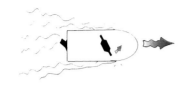

The jet drive relies totally on the waterjet to steer. No throttle - no steering.

A burst of throttle will turn the craft with little forward momentum. There's no clutch or gearbox.

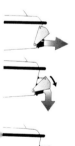

Some PW's have a clamshell arrangement, so they can reverse by deflecting the jet.
This is not a brake, you'll rip it off if you try to use it like one.
Use reverse only at low speeds.
The clamshell may also have a neutral position.

Some PW's have a variable trim mechanism.

Trim down will get you on the plane quicker, especially if you have a passenger.

Trim up will lift the nose, useful in choppy water.

The PW relies on water resistance to stop. Come off the power at speed and see how long it takes the machine to come to a halt. Try throttling down at various speeds - you'll need to know how long it takes to stop when approaching a beach where you will have to stop the engine before reaching the shallows. The PW can be stopped quickly by a radical change of course.
If something appears in front of you, a hard turn with full throttle will flip the machine out of the way. Coming off the power will simply mean you run into the object. Remember, you need throttle for thrust and steering.
If you are unfamiliar with your PW, take your first few rides in a quiet area. Practice turns, capsizing and recovery in water close to the shore.

Remember, you need throttle for thrust and steering. Coming off the power will simply mean you run into the object.

CAPSIZE

PWs are designed to turn over, it can be part of the fun. Always keep your buoyancy aid done up with the kill cord attached to your wrist or a strong point on your buoyancy aid.

After capsize

Make sure engine has cut out.

Check label on stern for way to rotate craft. You can damage the engine if you rotate it the wrong way.

Swim alongside, put one hand on grill and one on grab-handle. Put toe (or knee) on rubbing strake if possible.

Roll the craft over so the water drains off. Move to the stern, climb over back, do not tread on any of the jet parts, they are prone to damage.

Keep body low as craft is unstable, especially in choppy water.
Re-attach kill cord . Restart without choke but with slight throttle.

CAPSIZE WITH PASSENGER

Check the passenger is still with you and uninjured.

Keep passenger in sight, roll craft upright. Passenger should hold craft and help keep it steady.

Climb over the stern as before, assume driving position. Once seated and balanced, ask passenger to follow.

Restart machine

If engine does not restart the carbs may be flooded with fuel - leave it for a few minutes.
If it still won't start don't try to effect repairs out on the water.

The open engine hatch will unseal the machine, and it could be sunk by a capsize.
Get a tow, but first make sure you agree there is no fee. Stay with the machine. Don't let it drift. Someone could tow it in and claim salvage.

RULES OF THE ROAD

Most PW accidents are caused by collisions with other vessels, often PWs themselves. It is vitally important to know what to do if a collision looks likely. Try to make your reactions instinctive and always remember that the other driver may not have a clue. Assume it's up to you to avoid a collision.

FIRST RULE OF THE ROAD - KEEP A PROPER LOOKOUT
MOST COLLISIONS ARE CAUSED BY FAILURE TO SEE AN APPROACHING VESSEL IN TIME

Priorities
You will probably be faster and more manoeuvrable than anything else afloat. The rules generally require more manoeuvrable boats to keep clear of less manoeuvrable ones.

YOU MUST GIVE A WIDE BERTH TO

Dredgers and other working craft which are unable to manoeuvre

Fishing boats
Even if they don't appear to be fishing they could have booms out each side, or submerged tackle. They are experts at the sudden change of course when recovering gear.

Remember - keep an all round lookout, don't just look at the water ahead.

RULES OF THE ROAD

COURTESY AND COMMON SENSE

You should keep clear of:

Canoeists and rowers

Fishing boats are experts at the sudden change of course when recovering gear.

Large vessels, especially ferries
Supercats are massive versions of your PW. Don't play chicken with them - if you fall off in the turbulent wave and get sucked into their intakes, you're human puree.

Sailing boats they can only go where the wind will let them.

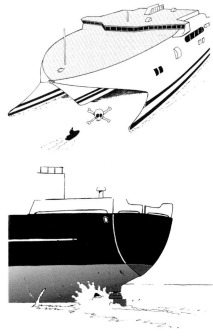

Boats operating with divers
Divers often surface some distance from their mother ship. Look for the international signal of a white and blue flag.

Stay well away from all commercial shipping. They can't stop their propellers nearly quick enough - and that's if they even see you. Never pass close to their bows or props, however daring you feel.

Promoting

The RYA is the national organisation which represents the interests of everyone who goes boating for pleasure.

The greater the membership, the louder our voice when it comes to protecting members' interests.

Apply for membership today, and support the RYA, to help the RYA support you.

nd Protecting Boating

Benefits of Membership

- Access to expert advice on all aspects of boating from legal wrangles to training matters

- Special members' discounts on a range of products and services including boat insurance, books, videos and class certificates

- Free issue of certificates of competence, increasingly asked for by everyone from overseas governments to holiday companies, insurance underwriters to boat hirers

- Access to the wide range of RYA publications, including the quarterly magazine

- Third Party insurance for windsurfing members

- Free Internet access with RYA-Online

- Special discounts on AA membership

- Regular offers in RYA Magazine

- ...and much more

Join online at **www.rya.org.uk**
or use the form overleaf.

Visit the website for information, advice, member services and web shop.

If you have previously been a member and know your membership number please enter here ☐☐☐☐☐☐☐

When completed, please send this form to: RYA RYA House Ensign Way Hamble Southampton SO31 4YA

	Tick box	Cash/Chq.	DD	Please indicate your main area of interest	❏ Powerboat Racing
Family†		£44	£41	❏ Yacht Racing ❏ Dinghy Cruising	❏ Windsurfing
Personal		£28	£25	❏ Yacht Cruising ❏ Personal Watercraft	❏ Motor Boating
Under 21		£11	£11	❏ Dinghy Racing ❏ Inland Waterways	❏ Sportsboats and RIBs

These prices are valid until 30.10.03   † Family Membership = 2 adults plus any U21s all living at the same address.

For details of Life Membership and paying over the phone by Credit/Debit card, please call 0845 345 0374/5 or join online at www.rya.org.uk

PLEASE USE BLOCK CAPITALS

Title Forename Surname Date of Birth Male Female

1.
2.
3.
4.

Address

Town County Postcode

Home Phone No. Day Phone No.

Facsimile No. Mobile No.

Email Address

Signature _____ Date _____

Instructions to your Bank or Building Society to pay by Direct Debit

Please fill in the form and send to:
RYA RYA House Ensign Way Hamble Southampton SO31 4YA   Tel: 0845 345 0400

Name and full postal address of your Bank/Building Society
To The Manager
Address
Bank/Building Society
Postcode

Originator's Identification Number
| 9 | 5 | 5 | 2 | 1 | 3 |

Reference Number

Instruction to your Bank or Building Society
Please pay Royal Yachting Association Direct Debits from the account detailed in this instruction subject to the safeguards assured by The Direct Debit Guarantee. I understand that this instruction may remain with the Royal Yachting Association and, if so, details will be passed electronically to my Bank/Building Society.

Name(s) of Account Holder(s)

Bank/Building Society account number

Branch Sort Code

Signature(s)
Date

Banks and Building Societies may not accept Direct Debit Instructions for some types of account

---

OR YOU CAN PAY BY CHEQUE

Source Code
077

Cheque enclosed £

Made payable to the Royal Yachting Association

Office use only: Membership number allocated

COLLISION AVOIDANCE

Remember - right is right.
Give way to the right, turn to the right. Think - RIGHT

Head to Head

Always turn to the right if a collision looks likely. Make the turn EARLY and OBVIOUS. You can't signal with hands, otherwise you'll come off the throttle, but you can signal with your entire PW. If in doubt, you can slow down or stop.

Throughout the world, at sea and in the air, head on collisions are avoided by both craft turning to the right.

If you can't go right (restricted channel - marina entrance) stop and signal the other vessel through.

In open waters

You only have to turn right if a collision looks likely, otherwise you may cause problems by crossing someone's bow.

In narrow channels, always keep to the right (starboard) hand bank, irrespective of which way you are going.

Sound signals

Boats often use sound to signal their intentions. If you hear a boat hooting at you, don't be offended, they are just communicating.

—
One short blast.
I am altering course to starboard.

— —
Two short blasts.
I am altering course to port

— — —
Three short blasts.
My engines are going astern. (Often used by ferries and water-taxis backing out of a berth into a main channel)

— — — — —
Five short blasts.
Your intentions are not understood. In other words, **Look Out!**

Most PW are unable to make a sound signal in return.

CROSSING

This one always causes confusion, but just think of it as a roundabout where you have to give way to the right.

Crossing
He is crossing from right to left

You must GIVE WAY. Alter course to GO BEHIND him. Do it EARLY. Make it OBVIOUS. Remember the yield zone. If in doubt slow and stop

Crossing
He is crossing from left to right

Now he has to give way - but watch him - he might not know the rules!
Be prepared to slow down or stop if he holds his course or speed.
Don't turn to port (left), in this situation it could make matters worse.

Remember, it is down to each driver to avoid a collision, even if he has to break the rules to do it.

OVERTAKING

In open waters, you can overtake either side, but stay well clear. The other driver may not have seen or heard you, and could suddenly swing in front of you.

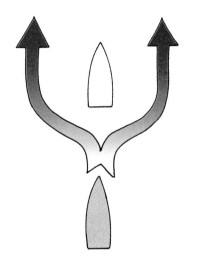

Use your head

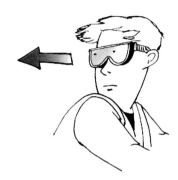

Always glance over your shoulder before altering course. Someone may be about to pass you. A good **ALL ROUND** lookout saves lives.

Don't trust the PW mirrors - they're normally covered in spray anyway. **USE YOUR HEAD.**

SPEED

In the open sea there is no precise speed limit in mph but the rules say that speed must be moderate.

Don't go fast near swimmers, where there are large groups of boats, in narrow channels, or close to beaches.

Watch out for speed limit signs in harbours.

WATERSKIING

Learn the internationally recognised hand signals to communicate with your skier.

Many multiple seater PW are capable of towing a skier but, as with any motor vessel, they have to follow the safety regulations, most of which are plain common sense.

All powerboats towing water skiers should be occupied by two competent persons, a driver and an observer. The driver can concentrate on the water ahead, whilst the observer can pass on the skier's signals.

Faster

Turn right

Turn around

Stop

Tow boats should be operated sensibly at a safe distance from people and property. Recklessness or deliberately endangering people is stupid and potentially fatal.

Water-skiing before dawn or after dusk is dangerous. Good visibility is required at all times.

When operating at sea or on a large expanse of water, the waterskier should wear a buoyancy aid. Keep well away from swimmers and swimming areas, which will be marked with yellow buoys.

Tow lines of floating line are normally not longer than 75ft.

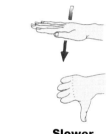
Slower

Stop   Skier down - stay clear.

Back to dock

I'm OK

INSHORE WATERS

Finding your way around.

Throughout Europe, the buoyage is positioned for boats to find their way INTO a harbour. Buoys that mark a well defined channel are called lateral marks, and are identified by shape and colour. Keep the can shaped reds (port hand markers) on your left and the cone-shaped greens (starboard hand marks) on your right when coming in.

Remember to check over your shoulder as you leave an unfamiliar harbour or beach. You'll have to find your way back again. Look for a conspicuous coastal landmark. Entrances to harbours are often hard to find. Memorise the way home.

It is often safer to navigate just outside the main channel to avoid large ships.

Points to remember

Buoyage takes you INTO a harbour. Reverse it for coming out.

Buoys or top marks are shaped so you can still identify them in poor light or glare.

Come home with the cones.

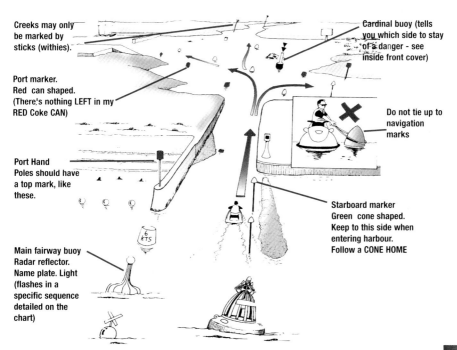

Creeks may only be marked by sticks (withies).

Port marker. Red can shaped. (There's nothing LEFT in my RED Coke CAN)

Port Hand Poles should have a top mark, like these.

Main fairway buoy Radar reflector. Name plate. Light (flashes in a specific sequence detailed on the chart)

Cardinal buoy (tells you which side to stay of a danger - see inside front cover)

Do not tie up to navigation marks

Starboard marker Green cone shaped. Keep to this side when entering harbour. Follow a CONE HOME

FURTHER OFFSHORE

If using your PW in an unfamiliar area or making a passage along the coast, you need to carry a chart. Study it before you go to see if there are any off-lying dangers. A PW is a boat, and just as vulnerable to coastal hazards as any other vessel. Treat the sea with respect.

A Small Craft Admiralty Chart, obtainable from most chandlers, can be wrapped in a special waterproof sleeve.

These details should include your intended route and destination and the time at which you expect to complete the expedition. Give your shore contact a call when you arrive or if you call off the expedition. Many hours of rescuer's time are wasted looking for people who are safe ashore but haven't made contact with relatives at home who are worried about them.

The weather might be fine when you set out but will it be like that all day?

A chart is a nautical map, showing all the things you would find on a land map and a lot more besides.

Carry a compass, such as a divers wrist type, to help you navigate.

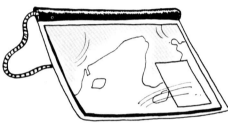

Fog is common off the coast at certain times of year, including high summer, and without a compass you can be totally lost within minutes.

Longer passages are best made in company with other boats or PWs which can provide support in the event of engine breakdowns or any other difficulties.

Leave details of your intended expedition with someone ashore.

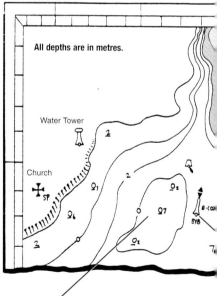

All depths are in metres.

This is a drying height, and shows an area that dries to a height above chart datum, the lowest possible low tide.

Check the forecast and make sure that the wind is not predicted to increase or the visibility to drop. PWs don't mix with very strong winds or fog, so be prepared to put off your plans until a more suitable day.

If you are going into unfamiliar waters, carry a compass and a chart and check on the tides and tidal streams for the day.

For longer trips a marine VHF radiotelephone is a very useful aid to safety. As well as allowing you to call for help in an emergency, it enables you to tell the Coastguard if you have taken shelter and are safe but unable to complete the planned trip.

Finally, before you set out, check that you have all the equipment you need and a good reserve of fuel - there are no petrol pumps at sea and you cannot walk to a filling station to buy an extra gallon if you run out.

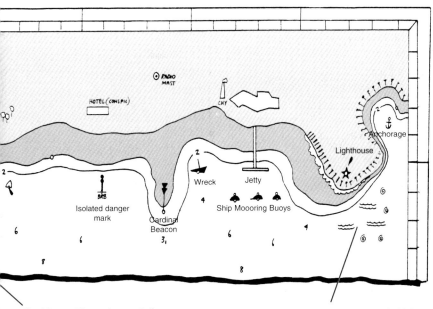

A cardinal buoy. These buoys tell you which side to stay (North, South, East or West) to avoid danger.

A headland. Off-lying shallows can create vicious overfalls.

23

TIDES AND TIDAL STREAMS

The UK and Atlantic coasts suffer from some quite impressive tides, unlike the almost non - tidal Mediterranean. In some areas, the difference between high and low water can be as little as 1 to 2 metres, but in others such as the Bristol Channel it can be as much as 12 metres. As a general rule, the tide rises and falls roughly once every twelve hours, six hours to flood and six hours to ebb. The time of high water advances by about 50 minutes each day. The height of the tide is governed by the phases of the moon. At Spring tides, which happen every two weeks, high tide will be higher and low tides lower; at neaps, also every two weeks, tidal differences are more moderate. What the tide is doing will affect how near the slip you park your car and where you launch and recover the PW.

As well as going up and down, the whole body of water moves sideways - six hours in one direction and six hours in the other. This is known as a tidal stream. In river estuaries and harbour entrances the incoming tide floods in and then ebbs out. In a narrow entrance, the tidal stream can reach speeds of up to 10 knots.

You can usually tell at a glance what the tidal stream is doing.

Boats moored to buoys will usually face into the tidal stream.
Buoys and other fixed objects will create a tell tale wake.
Be aware of tides in your area. A very low tide may make recovery difficult.

The tidal stream can carry a disabled PW away from an area very quickly possibly at speeds of up to 6 mph. Carry safety and survival gear on board at all times. (See page 31 coping with emergencies)

Tidal streams against wind can create short, sharp and confused seas.

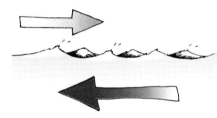

Wind with tide will have a calming effect.

Tide tables can be found at the harbour office, in nautical almanacs and at chandlers.

Reading a tide table
Tide tables show the times and heights of high and low water. The heights are given in metres above chart datum which is the lowest the water will ever go.

WALES – HOLYHEAD
Lat 53°19' N   Long 4°37' W

TIMES AND HEIGHTS OF HIGH AND LOW WATERS

	FEBRUARY				MARCH		
Time	m	Time	m	Time	m	Time	m
1 0339 0944 Sa 1603 2231	4·6 2·0 4·6 1·9	**16** 0531 1148 Su 1811	4·5 2·0 4·5	**1** 0206 0811 Sa 1426 2044	4·9 1·5 4·9 1·6	**16** 0321 0951 Su 1612) 2222	
2 0452 1100 Su 1723 2348	4·5 2·1 4·5 1·9	**17** 0025 0656 M 1308 1931	2·0 4·6 1·9 4·6	**2** 0256 0908 Su 1523 (2147	4·7 1·7 4·6 1·9	**17** 0441 1111 M 1733 2341	
3 0612 1220 M 1844	4·6 1·9 4·6	**18** 0136 0802 Tu 1411	1·9 4·8 1·7	**3** 0405 1022 M 1646	4·5 1·9 4·5	**18** 0611 1231 Tu 1901	

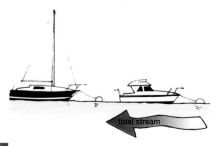

tidal stream

WIND AND WAVES

Personal watercraft can handle roughish seas, but you need to be an expert and it is very tiring. Should you be caught out in a squall, or if conditions deteriorate more than anticipated, try to read the waves. Throttle back at the crest otherwise the pump comes out of the water. Trim up to lift nose slightly. Fuel will be used much more quickly with constant throttle changes. Stand up PWs may find it easier to punch directly into the waves.

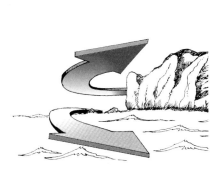

Both wind and tide accelerate around headlands. Expect to find rougher seas here

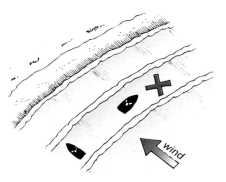

Avoid a lee shore

Attack waves at 40 degrees

Power on... off... and on again

**ROUGH WATER IS FUN,
BUT RE-BOARDING A CAPSIZED CRAFT
WILL BE PARTICULARLY DIFFICULT.**

LOCAL REGULATIONS

PWs have been targeted by many local authorities for special attention.

Under 16's may not be allowed to drive a PW unaccompanied. PWs are not toys, and require both physical maturity and experience of judgement which kids just don't have.

A number of local authorities will request a permit and insist that the craft carry a number on the bow.

Some busy commercial harbours have special channels to keep small craft out of the way of shipping. Some harbours may ban PWs altogether.

Many harbours have no-go areas usually to protect endangered wildlife. Stay away from these areas.

Look out for warning signs such as :-

10 knot speed limit

10 knot speed limit marker buoy

Personal watercraft permitted

Personal watercraft zone high visibilty marker

Personal watercraft activity prohibited

COURTESY TO OTHER WATER USERS

PWs have been criticised as a nuisance by other boat users. Consideration for other users will create greater harmony afloat.

Noise travels easily across water, especially in hot, still weather. Using a PW in one area all day will annoy others.

Don't weave in and out of an anchorage at high speed. Your tell-tale may end up in peoples' cockpits, or in their lunch.
You may also fail to spot people swimming around their boats. Keep the high speed stuff away from anchored boats.

Avoid areas where there are swimmers. If close inshore, drop to displacement speeds and keep a good lookout.

Don't rev-up in shallow beach areas - you will spray beach-users with sand and gravel.

Don't wave jump behind a boat. Fun though it is, the close proximity of a PW can be unnerving and irritating. There are plenty of waves to jump at sea. Stern hogging boats is a nuisance and dangerous.

Don't weave in and out of an anchorage

SOURCES OF WEATHER INFORMATION

Weather forecasts are available from many sources including radio, television and weather charts posted outside harbour offices.

The shipping forecast including the forecast for inshore waters is broadcast every day on Radio 4 at 0535 (FM and Longwave), 1201 (Longwave) 1754 (Longwave-weekends FM also) and 0048 (FM and Longwave). Wind speeds are given using the Beaufort scale.

THE PW BEAUFORT SCALE FOR OPEN WATERS.

Force 1: Wind speed 1-3 knots. Just ripples on a smooth sea. Perfect weather for a long distance run.

Force 2. Wind speed 4-6 knots. Small wavelets, not breaking.

Force 3. Wind speed 7-10 knots. Gentle breeze. Large wavelets, crests beginning to break. Wave height about 1 ft. Beginners will struggle.

Force 4. Wind speed 11-16 knots. Small waves growing longer. Fairly frequent white horses. Experienced hands only.

Force 5. Wind speed 17-21 knots. Fresh breeze. Moderate waves taking a more pronounced form. Many white horses, perhaps some spray. At the top end of PW driving. Hardened experts only. Very tiring and becoming very dangerous.

Force 6. Wind speed 22-27 knots. Strong breeze. Large waves forming. White foam. Crests more extensive. Forget it. Put the PW away for another day. You'll only break it.

Force 7 - 11. Definitely do something else.

Forecasts will also give an idea of visibility. Fog is the greatest threat to an offshore PW as they have no electronic navigation aids and drivers can easily become disorientated.

KNOTS, ANCHORING AND TOWING

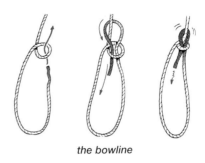

Knots
You only need to know two knots to use a PW.

Bowline - a good strong knot useful for towing or anchoring

the bowline

Round turn and two half hitches - to attach lines to the tow hitch for shallow launching (see page 11)

Anchoring
The anchor line attaches to the winch eye at the bow.
Using the small grapnel and a 5m length of rope, a PW can be safely anchored in shallow water. A careful eye needs to be kept on it, as a rising tide can make the anchor less effective. Some firms produce special bag anchors. Alternatively, a bag full of stones can be used.
Make sure the anchor is securely on the bottom, if it's not your PW will float away.

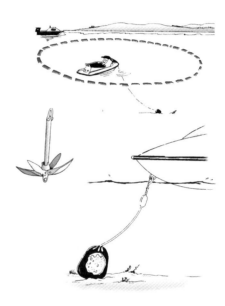

Towing
Use a bowline for towing a PW. Keep the towing speed to a minimum, no faster than a brisk walking pace, water can be forced into the engine if towing speed is too high. The owners' manual will tell you how to close off water intakes if a long tow is anticipated.

If a larger vessel comes to your aid, they may be able to take your PW aboard.
Before accepting help from another vessel, check that they are not going to make a claim for salvage against you. It's unlikely, but some individuals will offer you help and, when your PW is recovered, demand part of its' value, expecting your insurance company to pay up. Agree terms before you accept the tow-rope, even if it is only the promise of a drink once safely ashore.

RECOVERY FROM THE WATER

Recovering the PW is the reverse of launching. When approaching a beach, cut the engine before the water shallows to less than three feet to avoid engine damage. You may have to jump off and walk it in.

Put the trailer in the water.

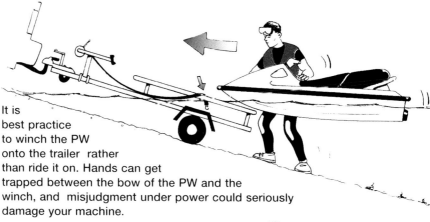

It is best practice to winch the PW onto the trailer rather than ride it on. Hands can get trapped between the bow of the PW and the winch, and misjudgment under power could seriously damage your machine.

Cut the engine and make sure the craft is securely attached to the trailer before pulling it out. Once clear of the water, attach a freshwater hose (if available) to the special flushing intake (see owners' manual) and run machine for a minute to purge the salt water from the system.

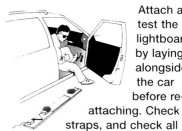

Attach and test the lightboard it by laying it alongside the car before re-attaching. Check straps, and check all seats and hatches are properly closed for the journey home.

If it's a residential area, this may be better done at home, as several PWs purging at once will create a nuisance. You don't want a petition going around to have the slipway closed to Personal Watercraft.

Wet suits and other gear can be washed in a mild detergent in the bath.

SAFETY AND EMERGENCIES

The safest way to go to sea is with another PW, so there is a spare craft to assist if you break down. Make sure you have a length of rope with you or the other craft won't be able to help.

When towing, attach the line to a proper towing point with a bowline. Your helm will provide some steering, but keep the speed fairly low. The towing vessel will use more fuel than usual.

If on your own, attracting the attention of other craft will be the biggest problem. If you are broken down but not in any immediate danger, try waving or beckoning to any passing boat.

Visual Distress Signals

In a more threatening situation, when you think you will be in grave danger and need immediate assistance you can use one of the internationally recognised distress signals:

Arm Movements. Up and down outstretched arms. (Note: Do this very slowly, as fast arm movements are little more than a blur at any distance.)

Flares. In bright sunlight the most effective signal is an orange smoke flare. In dull, overcast conditions a red hand-held flare is visible from further away. Don't use all your flares at once, keep one in reserve to pinpoint your position when you can see a lifeboat, helicopter or other potential rescuer.

If you are really in danger do not hesitate to call for help - delay could be fatal in deteriorating weather conditions or when night is falling. But you must not use a distress signal if all you want is a tow back to the shore on a calm day and there are plenty of other boats or PWs around who could help.

VHF Radiotelephones. A marine band VHF radiotelephone allows you to talk direct to the Coastguard and to other boats at sea.
If you are going on long trips it is worth having one but you will need a radiotelephone operators certificate and an annual licence for the set. Details of courses for the operators certificates are available from the RYA.

Mobile phone. Kept in a waterproof bag, the mobile phone can be used to raise help close to shore. Dialling 999 will raise the Coastguard, who are an emergency service. Be sure to tell them the nature of your emergency and exactly where you are because, unlike VHF transmissions, your position cannot be established from a mobile phone call.

Mobile phones were never designed to go to sea, don't put your faith entirely in them. Keep the battery fully charged.

UNACCEPTABLE RISKS

While highly manoeverable and versatile, PW are vulnerable to the same dangers as other craft.

Avoid fog. The PW will give a feeble radar return, if any, and could easily be run down.

Avoid travelling at night. Even with navigation lights, which any powered vessel must carry during the hours of darkness, the PW's small profile makes it vulnerable. A drifting PW will be even harder to spot.

Avoid breaking water and tidal rips. In these conditions a capsized PW could easily be swept onto rocks, or caught in undertows that will make reboarding impossible.

Riding whilst intoxicated is a short cut to disaster. Balance, vision and judgement are all affected, as is bravado. Alcohol and high performance boats of all types do not mix. Don't drink and ride.